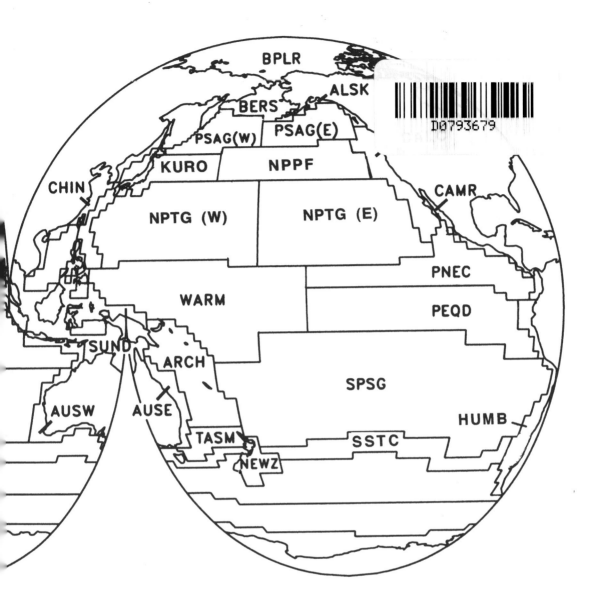

EAFR   Eastern Africa Coastal Province
INDE   Eastern India Coastal Province
INDW   Western India Coastal Province
REDS   Red Sea, Persian Gulf Province

INDIAN OCEAN TRADE WIND BIOME
ISSG   Indian South Subtropical Gyre Province
MONS   Indian Monsoon Gyres Province

PACIFIC COASTAL BIOME
ALSK   Alaska Downwelling Coastal Province
AUSE   East Australian Coastal Province
CALC   California Current Province
CAMR   Central American Coastal Province
CHIN   China Sea Coastal Province
HUMB   Humboldt Current Coastal Province
NEWZ   New Zealand Coastal Province
SUND   Sunda-Arafura Shelves Province

PACIFIC POLAR BIOME
BERS   North Pacific Epicontinental Sea Province

PACIFIC TRADE WIND BIOME
ARCH   Archipelagic Deep Basins Province
NPTG   North Pacific Tropical Gyre Province
PEQD   Pacific Equatorial Divergence Province
PNEC   North Pacific Equatorial Countercurrent
        Province
SPSG   South Pacific Subtropical Gyre Province
WARM   Western Pacific Warm Pool Province

PACIFIC WESTERLY WINDS BIOME
KURO   Kuroshio Current Province
NPPF   North Pacific Transition Zone Province
PSAG   Pacific Subarctic Gyres (East and West)
        Province
TASM   Tasman Sea Province

Ecological
Geography
of the
Sea

Cover photograph: © Jim Brandenburg, Minden Pictures

This book is printed on acid-free paper. ∞

Copyright © 1998 by ACADEMIC PRESS

Academic Press
a division of Harcourt Brace & Company
525 B Street, Suite 1900, San Diego, California 92101-4495, USA
http://www.apnet.com

Academic Press Limited
24-28 Oval Road, London NW1 7DX, UK
http://www.hbuk.co.uk/ap/

Library of Congress Card Catalog Number: 98-84363

International Standard Book Number: 0-12-455558-6

PRINTED IN THE UNITED STATES OF AMERICA
98 99 00 01 02 03 MM 9 8 7 6 5 4 3 2 1

Ecological
Geography
of the
Sea

ALAN LONGHURST

ACADEMIC PRESS

San Diego   London   Boston   New York   Sydney   Tokyo   Toronto

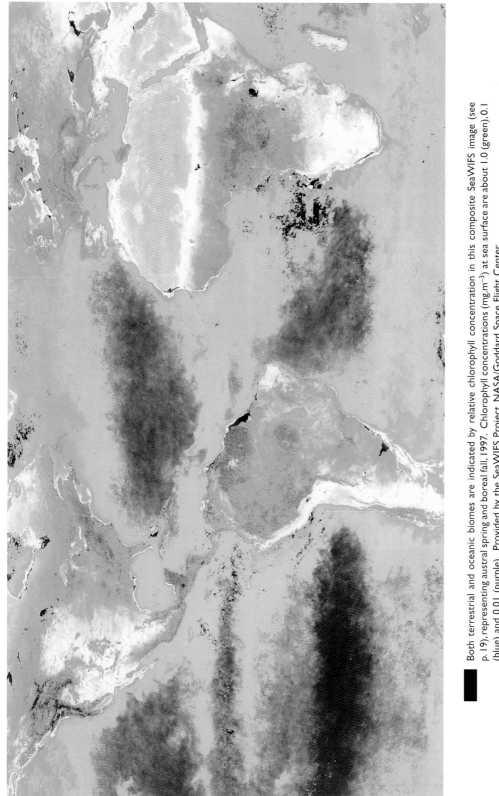

Both terrestrial and oceanic biomes are indicated by relative chlorophyll concentration in this composite SeaWIFS image (see p. 19), representing austral spring and boreal fall, 1997. Chlorophyll concentrations (mg.m^{-3}) at sea surface are about 1.0 (green), 0.1 (blue) and 0.01 (purple). Provided by the SeaWIFS Project, NASA/Goddard Space Flight Center.

"Thus, in these regions at least, observations give better results than theories can provide . . ."

Klaus Wyrtki, *Naga Report*, 1961

CONTENTS

7 The Atlantic Ocean 99

8 The Indian Ocean 205

9 The Pacific Ocean 245

10 The Southern Ocean   339

 **PREFACE**

How does the ecology of plankton respond to regional oceanography?

This book is an attempt to answer this apparently simple question, not by reference to the distribution of individual species but by analyzing how some basic ecological processes vary in response to the characteristics of regional oceanography. From this analysis, I have tried to draft an ecological geography of the oceans and shallow seas so that, at some level of probability, a prediction may be made about the characteristic ecology of any region.

The main part of the book is arranged geographically by individual ocean, as in classical biogeographies. I have chosen this arrangement because the reader must already have access to several modern texts on marine ecology and the physiology of marine organisms, arranged so as to make it easy to extract information about individual ecological or physiological processes. On the other hand, it may be quite difficult to locate organized information about a particular region of interest other than those few parts of the ocean that have attracted major expeditions in recent decades. Therefore, for the same reasons that Tomczak and Godfrey (1994) published their innovative text on regional oceanography for physicists who lacked a modern geography of the oceans, this book is offered as a geography for marine ecologists.

It will be obvious to the reader that I must have consulted many more original papers than I have been able to cite: This is inevitable in a work such as this, lest the bibliography grow as long as the text. I apologize in advance to those who spot their own uncited contributions to the whole. I have tried to list sufficient key papers to give access to the literature of each province rather than to support each of my contentions with a citation. You will also quickly come to appreciate that I have been unable to locate comprehensive ecological studies for several regions (usually distant from Europe or North America) even if their physical oceanography is quite well-known. This says something about the priorities of oceanographic agencies and also about how physical and biological oceanographers have been trained to look differently at the world.

During the past decade there has been a major cognitive shift among biological oceanographers from an assumption that a simple "nitrate–diatoms–copepods–fish" model adequately summarized the pelagic food chain to more complex constructs that incorporate picoautotrophs, mixotrophic protists, and the role of bacteria in regenerating nitrogen for autotrophs, to say nothing of a more realistic range of possible mechanisms for the flux of major nutrients: Mills (1989) suggests why it took so long to incorporate in our general model of the pelagos the observations of Lohman, made at the end of the nineteenth century, of the relative abundance of nanoplankton cells. Inevitably, most of the region-specific studies I have consulted were done when the simpler model was the accepted norm, and relatively few regional studies incorporate the current models. For some regions I have been able to suggest how the older studies should probably now be interpreted and have extrapolated accordingly, whereas for others I have simply described what I think is known and left you to guess the rest.

I am acutely conscious of the debt I owe to so many people (some now deceased) who taught me enough about the oceans to risk undertaking this book. They are far too numerous to list, but I remember especially the following: Gunnar Thorson, Theodore Monod, and Vernon Bainbridge got me started in the Gulf of Guinea in the 1950s; Bill Thomas and Carl Lorenzen introduced me to oceanic plankton during EASTROPAC in the 1960s; Maurice Blackburn and Ray Beverton showed me the close relationship between fish and temperature off Baja California and Spitzbergen, respectively; Radway Allen and Benny Schaefer helped me understand how fish populations worked; Reuben Lasker and Robert LeBorgne taught me something about physiological rate functions for zooplankton and nekton; Philip Radford introduced me to flux models of ecosystems; and with Bob Williams I explored vertical plankton profiles both in data archives and at sea. Warren Wooster, Karl Banse, and David Cushing always seemed to be there when I needed advice or encouragement.

In recent years, I was strongly supported at the Bedford Institute of Oceanography (Department of Fisheries and Oceans, Canada) by my colleagues of the Biological Oceanography Division with whom I worked at sea in the eastern tropical Pacific, in the Canadian arctic archipelago, and in the North Atlantic from 1977 until 1993. I thank them all for their patience in helping me using their varied expertise in pelagic ecology (Trevor Platt, Shubha Sathyendranath, Glen Harrison, Bill Li, Erica Head, Bob Conover, Paul Kepkay, and Doug Sameoto) and ocean physics (Ed Horne, Fred Dobson, John Lazier, and Allyn Clarke).

We sailed aboard *Hudson,* described to me not so long ago by the late academician Konstantin Fedorov as "the most capable oceanographic ship in the Western world." She is still the only surface ship to have circumnavigated the whole American continent, around Cape Horn, and through the Northwest

Passage in one voyage, but now she teeters on the edge of following our other Canadian research vessels *Baffin* and *Dawson* to scrap. They may be replaced by a Coast Guard hand-me-down.

Finally, this is a good place to apologize to Françoise for all those summers when I was at sea instead of with the family and for all those weekends at the computer.

INTRODUCTION

Ideally, pelagic biogeography should have three components. First, it should describe how, and suggest why, individual species from bacterioplankton to whales are distributed in all oceans and seas. Second, it should tell us how these species aggregate to form characteristic ecosystems, sustaining optimum biomass of each individual component under characteristic regional conditions of temperature, nutrients, and irradiance. Third, and most important for some purposes, it should document the actual areas within which each characteristic pelagic ecosystem may be expected to occur.

The current state of pelagic biogeography is unsatisfactory on all three counts. Biogeographers have concerned themselves almost exclusively with the first question, but a comprehensive account of how species are distributed in the ocean still remains far beyond their grasp. The second question—what are the characteristic ecosystems of the ocean?—has been investigated for almost 100 years and much progress has been made, but there has been little interest in locating the geographical boundaries between ecosystems, which is the third requirement. It is my hope that this book will go a little way toward marrying the work of ecologists, or biological oceanographers, with that of regional oceanographers. It is only by integrating an ecological component into regional oceanography that we can hope to predict the characteristics of the pelagic ecosystem of a given oceanic area.

The simple descriptive phase of biological oceanography occupied almost a full century. Beginning with the straightforward exploration of the marine biota of all oceans by the *Challenger* in 1872–1876, this phase culminated with the multiship, seasonal explorations of large oceanic regions—NORPAC, IIOE, EASTROPAC, EQUALANT, and others—in the middle of the twentieth century. Though some of the very early voyages, notably the Plankton Expedition of 1889, also laid the foundations of our understanding of the biology of oceanic plankton, research on ecological processes was principally pursued in coastal laboratories—especially Kiel, Plymouth, and Woods Hole—where quite rapidly a basic understanding of the cycle of pelagic production and consumption was achieved, a process well described by Mills (1989); as Karl Banse reminds me, a major catalyst in this progress was the introduction into biological oceanography during the 1950s of bulk methods for chlorophyll, seston, proteins, and carbon uptake rates so that samples could be taken that matched the bottle data of the physical and chemical oceanographers.

In recent decades, the detailed study of ecological processes has been taken to the open ocean. Rather than exploring species distributions, the objective of recent large-scale and multiship expeditions has been to quantify the rate constants for physiological processes across a wide range of oceanographic conditions. The catalyst for this change of emphasis was surely the development of solid-state electronics for underwater sensors, for shipboard laboratory equipment, and for data processing. Reliable and compact instruments were critical for biological oceanography especially because of previous difficulties in obtaining data from below the sea surface.

During the past 25 years, and parallel with equivalent progress in the other branches of oceanography, these instruments have delivered nothing less than a revolution in our understanding of ecological dynamics in the oceans and in our ability to archive and process vary large quantities of numerical data, especially concerning the physical environment of the pelagic biota. This new phase of process studies in biological oceanography has now been carried sufficiently far that we can imagine two-dimensional numerical models that simulate characteristic marine ecosystems: coastal upwelling, spring bloom, tropical open ocean, temperate shelf, and so on.

However, there is a catch. Our ability to specify the critical physiological processes under a range of ambient conditions has outstripped our knowledge of the distribution, abundance, and biomass of the organisms themselves. Our accumulated data describing the geographical distribution of plankton still contains insufficient information to partition two-dimensional ecological models among a series of compartments, each having a characteristic ecology, that together might integrate the whole surface of the ocean. Lacking this possibility, global biogeochemical models have usually integrated ecological processes as a parameterized continuum.

This was the situation until very recently. Now, however, as the result of another technological revolution in the natural sciences, the required information to compartmentalize biogeochemical models is at hand. Sensors carried on earth-orbiting satellites can now obtain data representing conditions on the surface of the oceans at very short time intervals and over a grid of many closely spaced points. The NIMBUS satellite carried the Coastal Zone Colour Scanner (CZCS) sensors that for several years obtained data representing the surface color of the oceans and thus, with some reservations, phytoplankton chlorophyll. Other ocean color satellites and other sensor suites are planned to follow or are already orbiting in an experimental mode. Operational and routine chlorophyll sensing should follow later in this decade.

The NIMBUS sensors gave biological oceanographers a tool of unprecedented power. Running through this book is a thread of interpretation of the CZCS data, suggesting how we may use such information to progress from our general knowledge of pelagic ecology to a first approximation of what we have hitherto lacked: comprehensive, region-specific descriptions of seasonal changes in ecological processes over the whole ocean comparable to those we already have for terrestrial ecosystems. In short, it outlines proposals for an ecological geography of the pelagic ecosystem of the surface waters of the ocean.

Why are the proposals restricted to the pelagos, and particularly of the open ocean? Simply because the new satellite sensors tell us nothing about the largest part of the marine biosphere—the interior of the ocean—or about the organisms of the seabed. So, despite their importance in mediating global carbon flux through the respiration of sinking organic material, the biota living below the surface layer remain difficult to study. The same must be said for the benthic communities for which there seems to be no substitute for the slow and steady accumulation of the observations that have occupied a small body of marine ecologists throughout this century. Therefore, you will find little about the benthos or bathypelagic ecology in this book. It is difficult to see how progress in understanding these ecosystems can accelerate in the foreseeable future.

Nor could the CZCS sensors resolve biological properties shorewards of the turbidity front that occurs along almost all coasts at some depth less than 50 m and where, in any event, the fractal nature of coastlines will continue to frustrate logical and comprehensive mapping of their ecological characteristics. Thus, I have had to be very cautious in what I have said about the neritic zone.

Philosophically, if we finally seem to have the right kind of information to define an ecological geography of the pelagic realm, then it should be attempted if only to test the quality of the new data. However, more practically, maps of the ecological characteristics of the ocean must bring aid and comfort to those who use biogeochemical models to quantify the effects of climate change on oceanic biota and, inversely, of the mediation of climate change by oceanic biota. Perhaps a geographic synthesis may also help those who monitor fisheries production to quantify the effects of environmental variability on renewable resources.

In this chapter I examine the principles that might be used to partition the pelagic ecosystem into objectively defined geographic compartments. It is far from evident how this can be done since ecological boundaries in the ocean (even if it is agreed that such boundaries exist) must change their location seasonally and between years. Therefore, how can they possibly be mapped? How can the coordinates of the boundaries be usefully defined? How can they accommodate the consequences of the seasonal meridional march of the radiation and wind fields or the effects of the El Niño–Southern Oscillation events that modify global wind fields and circulation patterns every few years? Even more, can they possibly capture the effects of longer, decadal-scale changes in weather patterns?

I cannot emphasize too strongly that the maps of biogeographic provinces offered in this study are not intended to represent boundaries that are fixed in space and time but are only intended to indicate the approximate time-averaged spatial relations between provinces. How to actually locate the boundaries between provinces when required for global integration of a set of data representing a real period of time is a separate and complex question. For this to be possible, the boundaries must be defined by features that can be observed by remote sensing, for example, discontinuities in the slope or elevation of the sea surface or in the sea-surface thermal or color fields. Only from earth-orbiting remote sensors can we hope to have a flow of regional and global data sufficiently comprehensive and timely to predict the coordinates of variable boundaries between ecological provinces when required.

First, however, we have to define the discontinuities which may subsequently be located, perhaps by reference to proxy information, in real time. To do this, the first question we should address is whether the state of knowledge about the distribution of individual species of plankton (the present corpus of classical biogeography) is sufficient to suggest where we should place discontinuities in an ocean biogeography, and even whether such boundaries exist.

THE INADEQUACY OF CLASSICAL BIOGEOGRAPHY IN ECOLOGICAL ANALYSIS

I have already suggested that the results of 150 years of study of the distribution of the marine flora and fauna are so meager that they permit us to predict comprehensively the characteristic assemblage of species likely to occur in no region of the ocean. In fact, currently, the taxonomic biogeography of the sea belongs to the family of intractable scientific problems. I am not alone in this opinion: Dunbar (1979) and Rosen (1975) have commented that biogeography is an uncritical branch of marine science with neither a useful factual basis nor an agreed methodology: Dunbar stated, "The biogeographic method does not exist, or there are as many methods as biogeographers" and Rosen noted, "Biogeographers of today follow their own preferred and sometimes bizarre premises without casting more than a troubled glance at . . . conflicting theories and ideas." What follows tends to support such criticism, especially for oceanic biogeography, which has even more serious inherent difficulties than terrestrial biogeography (e.g., see de Beaufort, 1951): the cost of collecting samples on the high seas, the relative lack of isolation between natural regions, the high levels of expatriation among plankton species, and three-dimensional distributions that vary in both space and time. These are among the most serious problems, but not the only ones.

Despite the general difficulty of the discipline, terrestrial biogeographers quickly located the important discontinuities between the great faunistic divisions on land, but marine biogeographers still have only relatively sketchy accounts of faunistic or floristic discontinuities in the ocean. What came to be known as "Wallace's line" in the Indo-Pacific archipelago, separating the Australian from the Oriental terrestrial faunas, was identified by Alfred Russel Wallace in almost the same year that the *Challenger* set her sails for the first global exploration of the still-unknown marine biota. The general outlines of the taxonomy and biogeography of the terrestrial flora and fauna were sufficiently well described by the first half of the twentieth century that integration was already achieved between biogeographic regions (inhabited by organisms related in the taxonomic sense) and their constituent biotopes (inhabited by organisms forming characteristic ecosystems) as described by Pitelka (1941). In this way, it was possible to recognize ecologically similar but taxonomically distinct biotopes (e.g., see de Beaufort, 1951): for instance, the rain forests of Brazil, Papua-New Guinea, and Zaire are ecologically quite similar but taxonomically very different. However, we are nowhere near being able to make the same kind of analysis from the accumulated knowledge of the taxonomy and biogeography of oceanic biota, both aspects of which are still explored only very superficially.

This situation is paradoxical because there are probably more than 1.0×10^6 species of terrestrial animals, but no more than 1.0×10^4 species of marine plankton and nekton in all the animal phyla. Since the earliest voyages, a small part of the total effort of oceanography has been devoted to biogeography and even as late as the 1960s strong teams, expressly devoted to this task, were being recruited at some major oceanographic institutions. Even now, however, 150 years after the *Challenger* voyage, the total number of species in each major group of pelagic organisms is not even approximately agreed on and we have descriptions of the seasonally variable distribution of no more than a small proportion of them. This is really not surprising: Although a quantitative comparison cannot be made, the number of stations at which marine plankton have been collected, identified, and enumerated must be orders of magnitude smaller than that for the terrestrial invertebrates, though the area of the oceans is more than twice that of the continents.

It has been remarked that the basic unit of biogeography is the distribution of individual species, though even this simple adage is questionable because few taxonomists of the plankton have faced up to the relationship between binomial Linnaean "species" distinguished simply by some morphological criterion and the individual self-sustaining populations that occupy individual home ranges, are presumed to have some genetic isolation, and may (or may not) be distinguishable on the sorting tray of a microscope. The surface of such problems has scarcely been scratched.

There is some agreement among taxonomists that the number of species (however defined) of mesozooplankton is around 2000 so that McGowan was able to write in 1971 that "holoplanktonic zooplankton are taxonomically well-known at the species level." In retrospect, this was surely an overoptimistic statement. Consider the taxonomic status of copepods of the genus *Calanus* during the ecological exploration of the oceans of recent decades: Two North Atlantic copepods of the genus are surely the best known of all zooplankton species, but *C. finmarchicus* (Gunnerus 1770) and *C. helgolandicus* (Claus 1863) were not clearly distinguished until the work of Fleminger and Hulseman (1977) and were, for a long time, thought to be geographical races (or subspecies) of a single cosmopolitan species. Recent work (quoted in Bucklin *et al.,* 1995) suggests that there are 14 species of *Calanus,* of which 3 (*C. hyperboreus, C. simillimus,* and *C. propinquus*) are morphologically distinct from the remainder, which are themselves distinguishable only by expert analysis of fine differences in secondary sexual characters of the exoskeleton. These comprise two species groups: a northern hemisphere arctoboreal trio (*C. finmarchicus,*

C. glacialis, and *C. marshallae*) and a larger group occupying midlatitudes in both hemispheres (*C. helgolandicus, C. pacificus, C. australis, C. orientalis, C. euxinus, C. agulhensis, C. chilensis,* and *C. sinicus*). Recently, molecular systematics using the DNA base sequences of the mitochondrial 16S rRNA gene (Bucklin *et al.,* 1995) has been used to confirm the reality of the group *C. finmarchicus + glacialis + marshallae* and also of the more genetically diverse *C. helgolandicus* group as listed previously, though genetic information on *C. orientalis* is still lacking.

However, this analysis also identifies significant differences between the individuals of *C. pacificus* from the California coast, from OWS "PAPA," and from Puget Sound so that "subspecies" level is accorded to individuals from these three locations, thus recalling Fleminger and Hulseman (1977), who referred to morphologically distinguishable forms of *C. helgolandicus* from the eastern and western Atlantic. In neither case do we know if the characters— morphological or genetic—lie along gradients or are discontinuous, and we understand very little about how species of *Calanus* may be sympatric. This unhappy situation will be found to extend to other important zooplankton genera: A similar molecular analysis of the difficult but abundant genus *Pseudocalanus* is currently in preparation by the same team.

I have reviewed this case in perhaps greater detail than it warrants, except as an example of how the present state of plankton taxonomy is a hindrance to ecological geography. However, this case is not the only one, and another indication of how very far we are from understanding the geography of pelagic species is given by recent ribosomal RNA analysis of 41 specimens of the small mesopelagic fish *Cyclothone alba,* which is ubiquitous and abundant in all subtropical and tropical oceans (Miya and Nishida, 1997). This analysis identified five monophyletic populations with low levels of mutual gene flow under conditions in which there appears to be no discernible barriers to prevent complete dispersion and intermingling of stocks: The central North Pacific population is genetically closer to those of the North Atlantic than to the other three populations inhabiting the Indo-Pacific ocean. Miya and Nishada reach the same conclusion as Dunbar (1979) and Rosen (1975): They write "we are a long way from knowing what species really exist in the oceanic pelagic realm."

Unfortunately, because RNA procedures are expensive and highly specialized, genome analysis will be very slow in helping us interpret the existing base of knowledge and leaves the biogeographic literature of many important groups of species still in rags and tatters and open to different interpretations. The critical reader of this book will quickly realize that in most cases in which the differential distribution of "species" of plankton is discussed, I have been quite unable to follow the taxonomic trail between the quotation of a specific name by a specific author and how the same specimens would be classified today.

Consider also the following statement (Dodge and Marshall, 1994) concerning the dinoflagellate genus *Ceratium:*

> Since the earlier work of Jorgenson . . . taxonomy within the genus has been fairly consistent and species can be readily determined. To date just over 100 species have been described, but there are problems with the large number of varieties which have been recognized although, over the years, many of these have been raised to species rank.

One might be forgiven for questioning the validity of the authors next comment: "For the preceding reasons *Ceratium* would seem an excellent, if not the best, dinoflagellate to use for biogeographic study"!

However, the identity of species in such well-known genera, though fertile ground for wrangling, is only the tip of the iceberg of this problem. Shih (1979) reminds us of the case of the thecosome mollusc *Cavolinia tridentata,* which exists as nine morphological types. Shih comments that such phenotypes may be allotted to species, subspecies, or forma according to apparently arbitrary decisions. Since such types also commonly appear to be arranged along a latitudinal (or thermal) gradient, it is not surprising that there often appear to be latitudinal discontinuities in the distribution of taxa. Between oceans, populations of apparently the same species frequently show morphological differences that to a future taxonomist may well suggest specific distinction. Shih cites the example of two species of *Rhincalanus*—one described from the Atlantic and one from the Indo-Pacific—of which the males are indistinguishable though there are minor differences between females from the two oceans: These have variously been considered to be two separate species or two forms of a single species.

Phenotypic variations in many phyla of plankton are the rule rather than the exception: Successive generations during a season may be morphologically distinguishable, nonreproducing expatriates may be different from individuals in the home range, individuals from warmer and colder, shoaler and deeper environments may differ, and so on. No wonder that five geographical forms of a common euphausiid *(Stylocheiron affine)* have been identified from the Indo-Pacific.

If one adds to such taxonomic problems those arising from the three-dimensional transport of biota, then certainty becomes stretched very thin indeed. Not only are boundaries between currents or water masses leaky, but the surface layers in which we are interested lie above deeper water masses that may have been subducted from the surface at convergent fronts even thousands of kilometers distant, together with at least some of their planktonic organisms. In this way, the characteristic surface organisms of a subtropical water mass may lie only a few hundred meters above organisms expatriated from a subarctic environment. In a simple plankton tow (and the vast majority of biogeographical data are obtained with such tows) organisms from the two environments may appear with great regularity in the same sample. About ½ km below the tropical copepods in the EASTROPAC zooplankton profiles, I often found quite large numbers of *Eucalanus bungii,* a copepod of the subarctic Northeast Pacific obviously subducted at the subarctic front; they were translucent, inactive, nonreproductive, and existed—one might think—only to confuse the biogeographer. Finally, we must remember the diel migrations which are performed by many species of mesozooplankton between the surface mixed layer and depths (300–500 m) at which we most commonly find expatriates from other water masses: These "interzonal" species are just that—they spend day and night in different depth zones and even different water masses.

Despite all these difficulties, Sinclair (1988) has reviewed the evidence for the existence of distinct, self-sustaining populations of oceanic zooplankton and is able to list about 20 studies which he believes successfully identified the general location of such populations. In several cases, the individual populations inhabiting different areas of the Pacific Ocean are morphologically distinct and, though direct evidence is lacking, may be assumed to be genetically isolated. It is usually assumed that in the highly dispersive oceanic environment, such plankton species are able to maintain a self-sustaining population, and consistently close their life history cycle within a specifiable region of the open ocean, often by means of appropriate seasonal vertical migrations between two depth zones with opposing mean flows. Some possible examples of this strategy will be discussed in the special regional descriptions in later chapters.

If there was a sufficient commonality between their home ranges, the boundaries of the areas occupied by self-sustaining populations might serve our purpose. Unfortunately, the total accrued information on such boundaries is insufficient to do more than offer some support—sometimes for conclusions we may have reached by other methods. This situation is unlikely to change rapidly because progress in marine biogeography, and the accumulation of new data on the distribution of pelagic organisms, will surely remain slow. Systematic biogeography is now perceived as a "filling-in" activity rather than an innovative branch of marine science, and this perception is unlikely to change despite current concerns for loss of biodiversity and habitat degradation by pollution and other forms of exploitation.

Nor can we realistically expect that a technological innovation will replace the slow and steady approach, which has changed little since the early days of oceanographic exploration. Although some progress has been made in mechanizing the slow chore of sorting and counting plankton samples, this is limited mostly to speeding the transfer of data to electronic media. Instrumentation to replace net sampling by counting and identifying organisms *in situ* with electronic and optical sensors is very far from replacing the need for time-consuming deployment of nets and the employment of specialists to identify the captured organisms. The most that has been done is to deploy imaging equipment in towed instruments that can obtain images of the larger organisms, from which a limited level of taxonomic identification may be inferred.

For all these reasons, there are still very few cases in which it has been possible to survey large regions of the oceans with a station spacing and frequency sufficient to specify individual distributions of the hundreds of species of plankton, nekton, and fishes occurring there, together with their seasonal and interannual distributions.

The California Cooperative Fisheries Investigations (CalCOFI) surveys are perhaps closest to the required model, having covered the California Current from about 1950 to the present day. There are two lessons to be learned from CalCOFI, however: (i) even though supported by both federal and state agencies (note, of the richest state in the richest union), the CalCOFI teams have been able fully to enumerate the plankton organisms from the surveys of only a single year and for the remaining years have had to be content with bulk measurements from the samples and accurate counts of a few selected taxa, mostly fish eggs and larvae; and (ii) even here the surveys have frequently been threatened with closure and have had to be performed on a progressively reduced frequency—currently only every third year. The other major large-scale survey of pelagic biogeography, the Continuous Plankton Recorder (CPR) surveys of the North Atlantic (e.g., Colebrook, 1982; Warner and Hay, 1994) along shipping routes from New England to Iceland and the North Sea, has also obtained samples since about 1950 and is almost the sole source of data on long-term changes in species distributions in the open ocean. These surveys, however, have been hard to maintain and currently have no assured long-term funding, being relegated to the status of a private foundation.

Some progress, of course, has been made in the analysis of how species distributions are associated with natural features of the ocean environment. The results are probably more reliable and useful for our purposes when a relatively small region has been analyzed rather carefully: A good example of this kind of approach is use of planktonic foraminifera as water mass indicators along a meridional transect from 30–55°N in the central North Atlantic (Ottens, 1991). Statistical species clustering of surface-collected foraminifera showed four distinct faunal assemblages associated with subpolar water (No. 1), the North Atlantic Current (No. 2), and the Azores Current (No. 4), with an additional assemblage (No. 3) lying zonally in the transition between the latter pair. These fit very well with the boundaries proposed for three ecological provinces in Chapter 7: Atlantic Subarctic Province is inhabited by No. 1, North Atlantic Drift Province by Nos. 2 and 3, and North Atlantic Subtropical Gyral Province by No. 4. Rather, I should have written that those small parts of these three provinces that were crossed by the transect were inhabited by these assemblages at one season of 1 year: I am not confident that we should extrapolate from this single transect to three wide zonal provinces of the North Atlantic and make assumptions about their characteristic assemblages of foraminifera.

There are very few analyses as promising as this one, but despite the tiny proportion of the surface of the ocean (or of the diversity of its biota) that has been included in these studies, attempts have been made since the earliest days to map biogeographical provinces in all oceans. These maps tend to have two characteristics in common: (i) The provinces do not have contiguous boundaries and (ii) the provinces resemble a series of latitudinal zones from poles to tropics, stretching across the ocean basins. Even though they usually conform to this general arrangement, there will be significant differences between almost any two examples that you might choose to compare.

Reviews of historical progress in biogeography generally start with the map of Steuer (1933) that is based on copepod distributions; this divided the oceans into circumpolar arctic, antarctic, and subantarctic regions and then divided each ocean basin (where appropriate) into subarctic and northern and southern subtropical and tropical regions. This is not very different from Sven Ekman's proposal at approximately the same time for a "pelagic warm-water fauna, northern and southern cold-water plankton, and a neritic plankton." From here to Beklemishev's (1969) map is not a far step, though he uses 20 instead of 11 regions. However, when compared, the two maps show limited congruence between the coordinates of comparable regions even in the same oceans. The most extensive compilation of such distribution maps is the comparative atlas of zooplankton of van der Spoel and Heymen (1983), but this does little more than remind us how far we are from achieving a comprehensive, species-based geography of the pelagic ecosystem and how little we have progressed since the

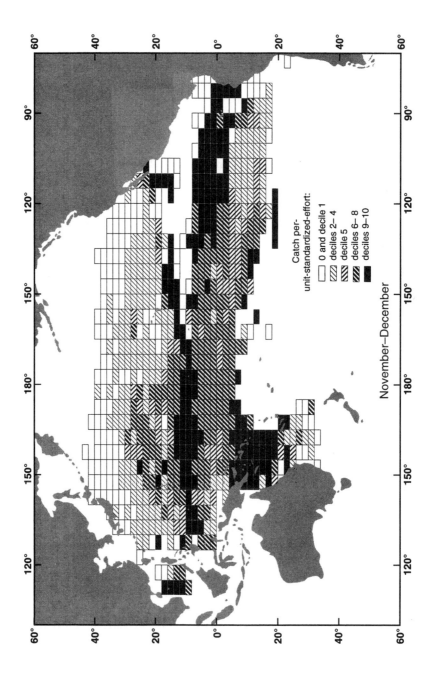

Catch per-
unit-standardized-effort:

0 and decile 1
deciles 2– 4
decile 5
deciles 6– 8
deciles 9–10

November–December

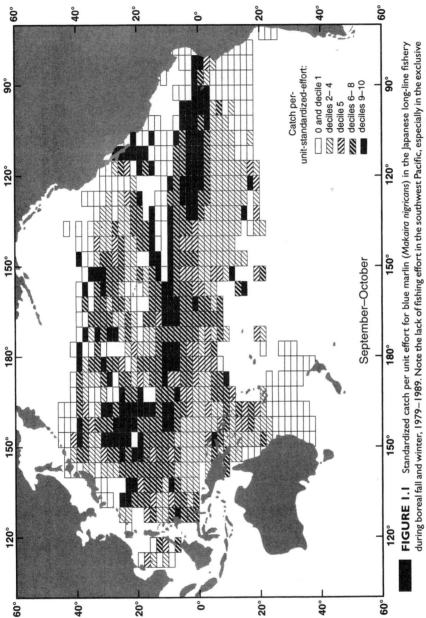

FIGURE 1.1 Standardized catch per unit effort for blue marlin (*Makaira nigricans*) in the Japanese long-line fishery during boreal fall and winter, 1979–1989. Note the lack of fishing effort in the southwest Pacific, especially in the exclusive economic zones of the Federated States of Micronesia and neighboring administrations, though adjacent catch rates suggest that marlin are nevertheless abundant there (from Hinton and Nakano, 1996, Fig. 5, with kind permission from Inter-American Tropical Tuna Commission).

Catch per-
unit-standardized-effort:

☐ 0 and decile 1
▨ deciles 2–4
▧ decile 5
▨ deciles 6–8
■ deciles 9–10

September–October

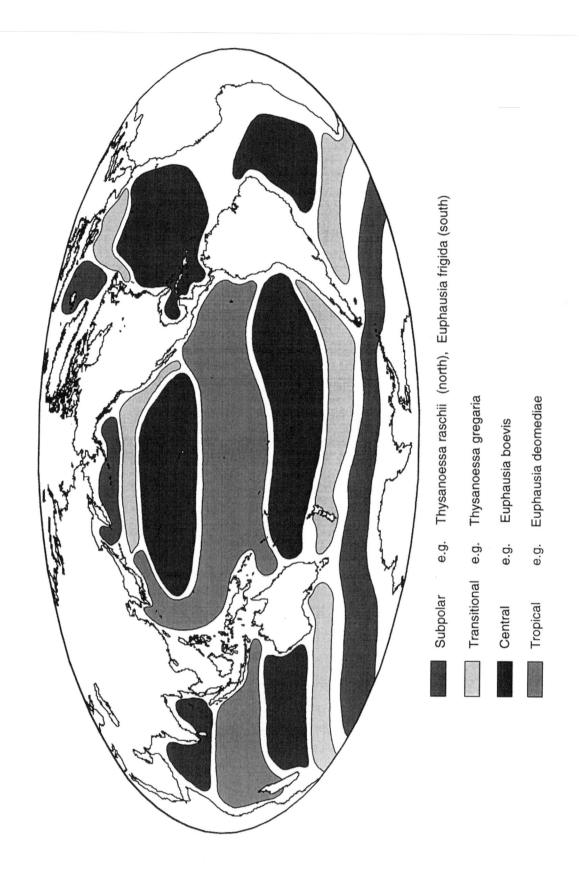

Subpolar   e.g.   Thysanoessa raschii (north), Euphausia frigida (south)

Transitional e.g.   Thysanoessa gregaria

Central   e.g.   Euphausia boevis

Tropical   e.g.   Euphausia deomediae

carly maps. After reviewing the approximately 60 maps that illustrate this atlas, I find it difficult to see any commonality between the distributions of different species.

Things appear even more confusing when proposals are made for partitioning the ocean into geographic units based on the distribution of a single group of organisms or even of single species; in this case, so many are useless or unconvincing because the author fails to show clearly, or even at all, the grid of samples on which the map is based from which we could see where a taxon occurred and where it is thought to be absent. It is worth noting how common this problem is, sometimes for quite unexpected reasons such as the exclusion of the sampling pattern from a particular zone of national jurisdiction (Fig. 1.1). Many (perhaps even most) published maps of species distributions show only those stations where the species was encountered in the samples; it is very common that the stations where samples were taken but the species was not found are unrecorded, which renders the map almost valueless. This kind of intellectual lapse reminds one that biogeography has been a very uncritical subject.

Of all the published maps of oceanwide distributions, few match those of Pacific euphausiids of Brinton (1962), the Pacific species distribution envelopes of McGowan (1971), or the maps from the North Atlantic CPR surveys (Anonymous, 1973). One is convinced by these presentations of the reality of some recurrent patterns of species distributions in the Pacific, the subarctic, central, equatorial, and eastern tropical species, together with those specializing in the transition zones at about 40° of latitude (Fig. 1.2); and in the North Atlantic, neritic, northeast intermediate, northeast oceanic, northwest oceanic, western intermediate, southern oceanic, and southeastern intermediate (Fig. 1.3). These maps provide information which will serve to confirm the reality of some of the ecological boundaries proposed in this work. Otherwise, the more synthetic global biogeographic maps (such as those of Beklemishev) will serve to indicate changes in the pelagic species of algae and zooplankton across some of the principal oceanographic discontinuities: I shall use some of these findings in Chapters 7–10 when I discuss the individual provinces. Such material is most useful in the southern ocean, in relation to the subtropical transition zones and perhaps at the boreal polar fronts. Apart from the neritic–oceanic transition, these appear to be the most important taxonomic discontinuities within the pelagos.

Several conclusions can be made from this brief review of the current state of oceanic biogeography. Most important, it is clear that the available description of the distribution of the species of the pelagic ecosystem is woefully inadequate to predict the distribution of the characteristic pelagic biomes which are the subject of this book. We shall therefore have to approach the matter differently.

FIGURE 1.2   The global distribution of groups of oceanic epipelagic euphausiids showing a general relationship with the biomes and provinces described in later chapters. The lack of a specialized group of tropical species in the Atlantic is attributed to the direct flux from the South Atlantic into the roots of Gulf Stream combined with the small zonal extent of the ocean. Groups of coastal species also occur, and match the meridional distribution of the oceanic species, but are not shown (derived from Brinton, 1962).

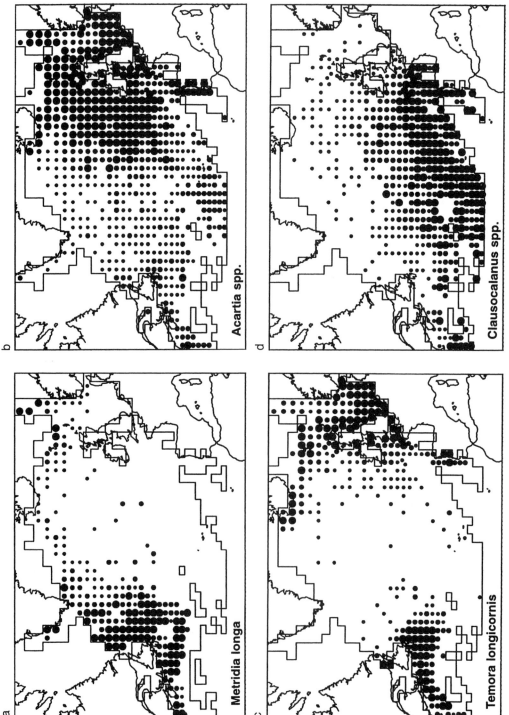

a   Metridia longa

b   Acartia spp.

c   Temora longicornis

d   Clausocalanus spp.

We may also concur that "pelagic biogeography has somehow not shared in the renaissance of interest that has occurred in the last decade for shallow-water and terrestrial systems" (summary report of the 1985 International Conference on Pelagic Biogeography), and at least one reason for this lack of energy seems obvious. Biogeographers could not participate in the solid-state electronic revolution of the 1970s that offered such intriguing new possibilities for biological oceanographers, who turned away from descriptive work to studies of ecological and physiological processes with tools that rapidly came to resemble those of physical oceanographers, at least in their ability to capture data.

It is only the exploration of the hitherto relatively unknown microplankton that has been greatly advanced by the new technology. Instruments transferred from the medical field, including laser optics in continuous flow cytometers and the use of fluorescence microscopy, have demonstrated the existence everywhere in the oceanic photic zone of novel micrometer-scale cells, the photosynthetic cyanobacteria and prochlorophytes. These cells are now known to be a significant component of the phytoplankton and critical in hitherto unsuspected ways for the flow of energy and material through the pelagic ecosystem.

During the same period, a conscious effort (Fasham, 1984) has been made to generalize critical physiological rates of algae and zooplankton. Agreement seems to be converging on values within a factor of less than 2, whereas prediction of regional numbers of organisms or their biomass could hardly be done with error terms of less than an order of magnitude. That this new ability to predict physiological rate functions characteristic of different environmental conditions should coincide with our new ability to measure at least one biomass index comprehensively at the surface of the oceans makes it all the more appropriate to revisit the question of an ecological geography of the ocean and to attempt to devise a scheme which will be valid for the entire surface of the ocean in all seasons and under all conditions.

THE NEW AVAILABILITY OF TIMELY, GLOBAL OCEANOGRAPHIC DATA

Novel data have become progressively available to oceanographers in the past 10 years from sensors carried on earth-orbiting satellites. Though only limited information is obtained at each data point on the ocean surface, the flow of simple data obtained every few days or even hours, and at a resolution of only a few kilometers over the whole surface of the ocean, has been revolutionary.

If these early developments mature, oceanographers of the twenty-first century will have everyday access to unprecedented information about the variability of ocean circulation and atmospheric forcing, and to a limited extent the future is already here. Though the sensors have been flown by government

FIGURE 1.3 Climatological distribution of four species of near-surface copepods in the Atlantic from a 60-year time series of Hardy Continuous Plankton Recorder samples from within the indicated area. (a) *Metridia longa* (western cold-water species); (b) *Acartia clausii* (northeastern intermediate species); (c) *Temora longicornis* (coastal species); (d) *Clausocalanus* spp. (southern oceanic species group) (courtesy of Plymouth Marine Laboratory).

agencies, frequently with military objectives, sufficient data have been made available to the oceanographic community as to open a new window on ocean physics. Primary users are rapidly starting to gain direct access to the incoming and stored data through the Internet, and this progress is unlikely to be reversed or halted. The data revolution which is in progress will certainly transform our ability to analyze regional oceanographic patterns, not only in physics but also in biology.

Real-time data on sea surface temperature are currently obtained from radiometers (AVHRR, VISRR, and VIRR) carried on earth-orbiting satellites and, in the coming decades, we can expect to have routine access to fields of surface winds (>13 existing or projected satellites through the end of this century), surface waves (7 satellites), and surface currents (14 existing or planned satellites). The success of TOPEX-POSEIDON 1992, which is currently measuring sea surface elevation with an accuracy of 2 or 3 cm and so is capable of mapping ocean circulation globally at short intervals, provides a glimpse of the future. One of the immediate and (relatively) unsophisticated benefits of the data from the new sensors has been an unprecedented ability to locate and map thermal fronts at sea down to the kilometer scale; in this way, there has already been a rapid increase in our knowledge of the locations of individual fronts, their evolution, and the physical processes that maintain them. Such information has been particularly valuable in understanding the nature of fronts at the shelf edge and also those associated with mesoscale eddies in the open ocean. By inference, and by comparison with chlorophyll images, the ecological significance of these features is now much better understood.

Near-surface chlorophyll fields for all oceans are also potentially available from satellite radiometers, based on the wavelength and intensity of backscattered visible light from the sea surface. Partial coverage of this field was obtained from 1978 to 1986 by the CZCS sensors carried aboard a NIMBUS satellite. Even at the first inspection, it was evident that the CZCS images contained entirely novel information concerning the global, seasonal distribution of phytoplankton chlorophyll. These distributions had previously been known only in broad outline from ship data. I shall frequently refer to the images of the global seasonal chlorophyll field which are shown in Color Plates 1–4.

The indicated chlorophyll values in the CZCS images represent not only surface chlorophyll but also integrate over a large fraction of the first optical attenuation depth for the relevant wavelengths, biased surfaceward where the chlorophyll profile is nonuniform. The sensed depth is approximated by $Z_e/4.6$, where Z_e is the euphotic depth (Morel and Berthon, 1989). Using the attenuation analysis of R. C. Smith (1981), we can infer that sensed depths range from about 25 m in clear oligotrophic water (0.1 mg chlorophyll m^{-3}) to about 5 m in eutrophic ocean water (10 mg chlorophyll m^{-3}).

Parenthetically, it will be convenient (in using these terms for the first time) to define how I shall use them. Originally referring to water containing low or high nutrient concentrations, "oligotrophic" and "eutrophic" are used more loosely these days respectively for (i) clear oceanic water perpetually with few biota and (usually but not always) low nutrients and (ii) green water with many biota and an originally abundant nutrient supply. If we can agree that "nutrient" means the Liebigian-limiting element (which may be a trace metal, as shall